First Library of Knowledge

Animals
of the World

BLACKBIRCH PRESS
An imprint of Thomson Gale, a part of The Thomson Corporation

Detroit • New York • San Francisco • San Diego • New Haven, Conn. • Waterville, Maine • London • Munich

First published in 2005 by Orpheus Books Ltd., 2 Church Green, Witney, Oxfordshire, OX28 4AW

First published in North America in 2006 by Thomson Gale

Copyright © 2005 Orpheus Books Ltd.

Created and produced: Rachel Coombs, Nicholas Harris, Sarah Harrison, Sarah Hartley, Emma Helbrough, Orpheus Books Ltd.

Text: Nicholas Harris

Consultant: Steve Parker

Illustrators: Mike Lowe, Stuart Carter, (*The Art Agency*), Sauro Giampaio

Other illustrators: Graham Austin, Andrew Beckett, Peter Dennis, Betti Ferrero, Giuliano Fornari, Ray Grinaway, Ian Jackson, Nicki Palin, Eric Robson, Peter David Scott, Colin Woolf, David Wright

For more information, contact
Blackbirch Press
27500 Drake Rd.
Farmington Hills, MI 48331-3535
Or you can visit our Internet site at http://www.gale.com

LIBRARY OF CONGRESS CATALOGING-IN-PUBLICATIONS

Harris, Nicholas, 1956-
Animals of the world / by Nicholas Harris.
p. cm. -- (First library of knowledge)
Originally published: Oxfordshire, UK : Orpheus Books, 2005.
Includes bibliographical references.
ISBN 1-4103-0348 9 (hardcover : alk. paper) 1. Animals--Juvenile literature. I. Title. II. Series.
QL49.H3245 2006
590--dc22   2005029624

Printed in Malaysia
10 9 8 7 6 5 4 3 2 1

CONTENTS

INTRODUCTION

ANIMALS are living things that can move about and must eat food to survive. This food may be plants, other animals, or both. Animals are divided into two groups: vertebrates (with backbones) and invertebrates (without backbones). Vertebrates include fish, **amphibians**, **reptiles**, birds, and mammals. Insects, spiders, worms, and mollusks are invertebrates.

FROGS

FROGS, toads, newts, and salamanders are all amphibians. They start their lives under water. When they grow up, they live on land. They return to the water to lay their eggs.

MOIST SKINS

Like most amphibians, frogs are cold-blooded, four-limbed animals. They have very small lungs, and they can take in air through their skins. They need to keep their skins moist to do this. So adults usually live near water.

FROG SPAWN

Frogs lay their eggs in ponds or streams. The eggs, or spawn, look like a clump of jelly. The black dots are the embryos. The babies that hatch look very different from their parents.

TADPOLES

The baby frogs, known as tadpoles, look like little fish with big heads and wriggly tails. Like fish, they breathe with gills. They feed on **microscopic** plants in the water.

After two months, the tadpole grows back legs and lungs. It loses its gills. To breathe, it comes to the water's surface.

A month later, its front limbs appear and its tail starts to shorten.

YOUNG ADULT

Four months after the young frog hatches, its eyes have grown bigger and its mouth is much wider. Instead of plants, it feeds on tiny creatures. About 0.4 inches (1 centimeter) long, the frog has lost its tail and is now ready for its adult life on land. It will be ready to breed in four years.

REPTILES

REPTILES are cold-blooded animals with scaly skins. Land-living reptiles must bask in the sun to warm up before going in search of food. Snakes, lizards, turtles, tortoises, and crocodiles are all reptiles. Most kinds lay eggs, but some give birth to live young.

Leatherback turtle

Matamata (river turtle)

REPTILES WITH SHELLS

Tortoises and turtles have shells. They are made of bony plates attached to the skeleton. The shell helps protect the animal from **predators**. Tortoises live on land. Some kinds can live for more than 100 years. Turtles live in rivers or the sea. They have lightweight shells and large flippers.

Giant tortoise

Anaconda

REPTILES WITHOUT LEGS

Indian cobra

Snakes can move quickly across the ground, swim, and even climb trees. All snakes are flesh eaters. Some, like the cobra, use their fangs to kill their prey with a venomous bite. Others, called constrictors, coil themselves around their victims and suffocate them.

The anaconda, from South America, grows up to 30 feet (9 meters) long. A constrictor, it swallows its victims whole with its loosely hinged jaws.

AGILE HUNTERS

Most lizards prey on small animals. If attacked themselves, some lizards, like geckos, can shed their tails and run away.

Dwarf gecko

TREE MONSTER

The green iguana, also from the South American rain forest, is a 7-foot-long (2-meter), tree-dwelling lizard. If danger threatens, it drops into the water below.

Green iguana

CROCODILES AND ALLIGATORS

CROCODILES and alligators are very powerful predators. They attack and eat other large animals—including humans. All these fierce reptiles have long, scaly bodies with thick, bony plates on their backs.

They live near rivers, lakes, and **estuaries** in tropical regions.

To tell an alligator from a crocodile, look at its teeth. An alligator's are not visible when its mouth is closed, but a crocodile's fourth tooth can be seen.

Crocodile hatchlings being carried in their mother's mouth

CARING FOR THEIR YOUNG

Crocodiles are very protective of their young. The female lays her eggs in a pit near the water's edge and covers them. She guards them from predators. After about three months, the hatchlings call out to her from inside the eggs. After they hatch, she carries her young inside her mouth to a safe area. There she looks after them for up to six months.

Young crocodiles feed on small prey. Adult crocodiles feed on large mammals that come to the water to drink. Seizing their prey in their jaws, the crocodiles pull them below water until they drown. Then they tear them apart with their teeth.

Young crocodiles eat crabs or frogs. Older ones feed on fish or animals already killed.

BIRDS

BIRDS are warm-blooded animals with four limbs. Their front limbs are wings. Their bodies are covered with feathers. They have toothless beaks. All birds lay hard-shelled eggs.

Toco toucan

The toucan uses its massive beak to grasp fruit.

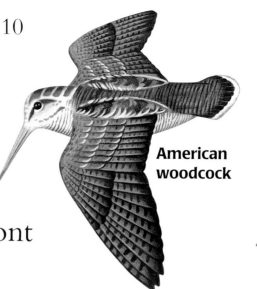

American woodcock

The woodcock uses its long bill to probe under the soil for worms.

WATER BIRDS

Many birds live close to rivers or lakes. There the food supply, including fish, insects and plants, is plentiful. Cranes are tall, long-legged birds. They inhabit wetlands and swamps. Flamingos get their pink color from the tiny plants and animals in the water on which they feed.

Greater flamingo

To feed, a flamingo holds its bill upside down in the water. Using its tongue, it pushes the water out through tiny comblike structures in its bill, leaving the food behind.

Siberian crane

The magpie, a type of crow, will eat almost any-thing.

Magpie

FLIGHTLESS BIRDS

Some birds cannot fly, but run or swim instead. Ostriches roam the dry grasslands of Africa in small groups. They use their excellent eyesight to spot predators. Their long, strong legs carry them quickly away from danger.

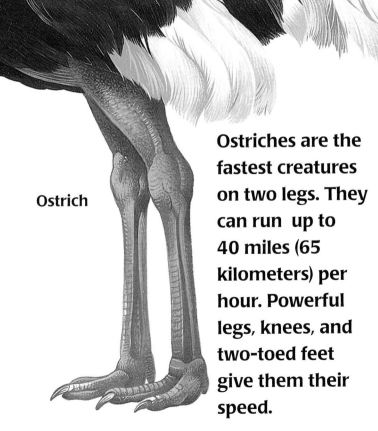

Ostrich

Ostriches are the fastest creatures on two legs. They can run up to 40 miles (65 kilometers) per hour. Powerful legs, knees, and two-toed feet give them their speed.

BIRDS OF THE NIGHT

Most birds find a safe place to rest at night. But owls hunt for prey—small mammals, insects and worms, fish, frogs, and other birds. Using its superb eyesight and hearing, an owl swoops down on its victim, seizing it in its razor-sharp talons (above). Its soft wing feathers (right) allow it to fly in almost complete silence.

Tawny owl

BIRD NESTS

BIRDS build nests to protect their eggs and their young from predators. The nests also keep the eggs warm while the adults sit over them. Birds build their nests in places that are hidden away, or out of reach of their enemies.

AN EAGLE'S NEST

Golden eagles make their nests high in trees or on rock ledges. Newly hatched baby eagles are covered in soft, fluffy feathers known as down. The hatchlings are fed on scraps brought by their parents. Gradually, they grow adult feathers.

BUILDING A NEST

Tailorbirds (left) sew the edges of leaves together to make a nest shaped like an envelope. The weaver is a skilled nest builder (right). It gathers up strips of leaves and weaves them into a deep pouch. The finished nest hangs from a branch.

Grebes build floating nests. Attackers must swim to reach them.

When ready to hatch, the baby eagle chips its way out of its shell using a special egg tooth on its beak.

FLIGHT

BIRDS, insects, and bats are the only animals that can fly. Birds have very light bones and strong wing muscles in their chests. They use their feather-covered wings to power themselves through the air.

Wandering albatross

The wandering albatross has the largest wingspan of any bird. It has extremely long, narrow wings. They are perfect for gliding and soaring in the winds over the Southern Ocean.

Mallard duck

Ducks are strong flyers. They can reach speeds of more than 62 miles (100 kilometers) per hour.

Racing pigeon

Spine-tailed swift

FAST FLYERS

Pigeons are among the fastest flyers. Their powerful wings can keep them airborne for long periods. Racing pigeons are experts at finding their way home over long distances.

Swifts are the fastest birds of all. Streamlined in shape, they swoop through the air catching insects in their gaping beaks. Some types of swift can fly at speeds of more than 93 miles (150 kilometers) per hour.

HOW BIRDS FLY

Like most birds, the scarlet macaw flaps its wings to push itself through the air. First, it lifts its wings high above its back (1). Then it pulls its wings downward and backward (2). This makes it move upward and forward. Then the wings start to move up again (3).

1

A scarlet macaw in flight

2

3

BIRDS THAT HOVER

Hummingbirds can beat their wings very quickly—as fast as 100 beats per second. This allows them to hover over flowers and collect nectar from them. They can also fly sideways or even backwards. Their wings beat so quickly they make a humming sound.

Andean condor

Hummingbird

The condor uses rising air currents, called thermals, to soar above the mountains.

WHAT IS A MAMMAL?

MAMMALS are warm-blooded animals. They care for their young and feed them with milk. Most mammals have four limbs and a covering of hair or fur.

Bats are the only mammals that can truly fly. Their wings are made of skin stretched between their long fingers.

Hog-nosed bat

CARING FOR THEIR YOUNG

All mammals, including humans, care for their young after they are born. While the young mammal feeds on its mother's milk, the mother protects it from predators. Some, like baby horses, can walk soon after birth, while others, like baby rabbits (or humans), are completely helpless.

Killer whale

Many sea mammals have little or no hair covering. To help them swim easily, seals have flippers instead of arms and legs. Whales and dolphins have no hind limbs at all.

Elephant seal

PLATYPUS

The young of a platypus, a river-dwelling mammal from Australia, are born inside soft-shelled eggs. They hatch out after a few days.

The platypus uses its bill to find its prey.

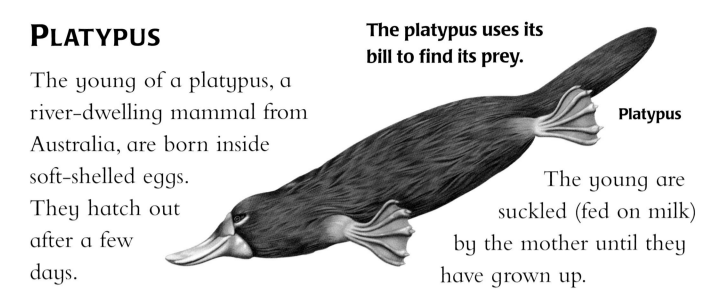

Platypus

The young are suckled (fed on milk) by the mother until they have grown up.

PREHISTORIC MAMMALS

Mammals are descended from reptiles. They first appeared more than 200 million years ago, when dinosaurs roamed the Earth. Different kinds of mammals lived in the past, including elephants with backward-pointing tusks, giant guinea pigs, short-necked giraffes, and miniature horses.

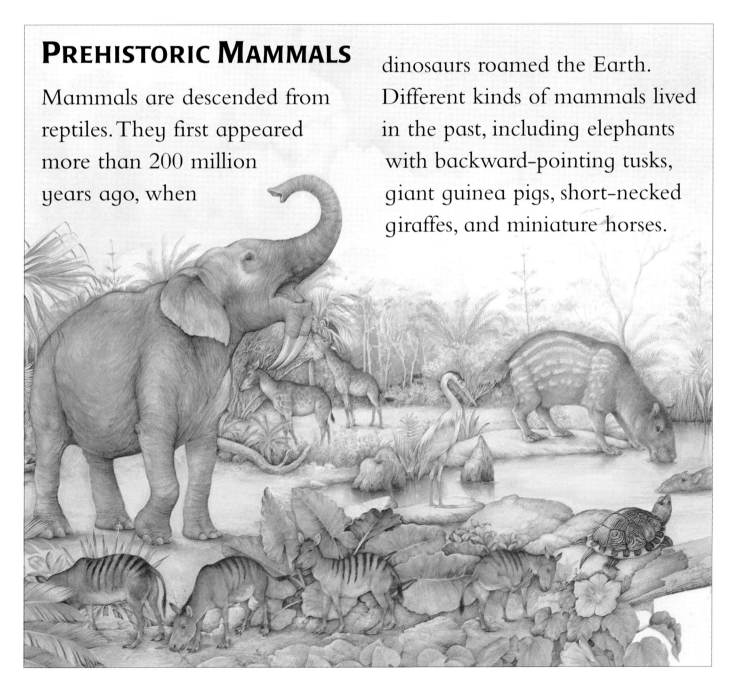

MAMMAL FAMILIES

THERE ARE about 5,000 different kinds of mammals. They range in size from tiny shrews, only a few inches long, to the 98-foot (30-meter) blue whale.

Three-toed sloth

Sloths, from the South American rain forest, spend most of their time hanging upside down in the trees.

Hare

GNAWING TEETH

All rodents, including mice, rats, porcupines, and squirrels, have long, continuously growing, sharp front teeth. Rabbits and hares also have them. Most of these animals eat only plants.

Brown rat

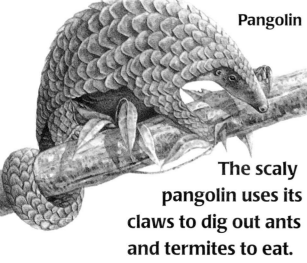

Pangolin

The scaly pangolin uses its claws to dig out ants and termites to eat.

White rhinoceros

The white rhinoceros grazes on the grassy plains of Africa. Its huge body is covered in thick skin.

Hoofed Mammals

Giraffes and rhinoceroses belong to the group of mammals that have hooves instead of claws on their feet. Giraffes are related to the deer family. They roam the African grasslands feeding on leaves and shoots from the tops of trees.

Giraffe

Standing over 16 feet (5 meters) in height, the giraffe is the tallest animal on land. It has excellent eyesight and can run over 31 miles (50 kilometers) per hour.

Giant panda

Endangered!

Both the giant panda and tiger from East Asia are now very rare in the wild. As the forests where they live are cut down, they have fewer places to go. Pandas are also very slow to breed.

Tiger

MARSUPIALS

LIKE other kinds of mammals, **marsupials** give birth to live young. But the babies are very tiny and have to be cared for inside the furry pouch on their mother's belly.

Apart from the opossum family of the Americas, marsupials live only in Australasia. Many feed only on plants, leaves, and fruit, but a few prey on insects and small animals.

KOALAS

Koalas spend nearly all their lives in the trees, stirring only for a few hours each night to feed on eucalyptus leaves and shoots. They use their grasping hands to cling to tree trunks—and to their mother's back!

Male kangaroos fighting

KANGAROOS

Kangaroos live in groups on the hot, dry plains of Australia. They can cover enormous distances bounding along on their strong back legs. Males fight one another to be the head of the herd. Females have deep pouches that are forward facing so that her young, called joeys, do not fall out.

OPOSSUM

The Virginia opossum is the only marsupial that lives in North America. It often searches for food in trash cans. When threatened, it may pretend to be dead. This is called "playing possum."

A female kangaroo's nipples are inside its pouch. The tiny baby sucks the nipple until it is old enough to let go.

Joey inside mother's pouch

HUNTING

Cheetah

SOME animals seek out and kill other animals for food. Hunters must use stealth, speed, and strength to overcome their prey.

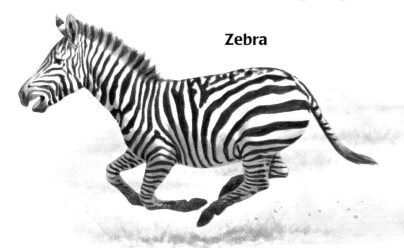

Zebra

SPRINT CHAMPION

A powerful hunter from the dry grasslands of Africa, the cheetah is the fastest animal on land. This big cat can run faster than 60 miles (97 kilometers) per hour.

INTO THE ATTACK

A cheetah may stalk its prey—in this case, a young zebra—for up to several hours. When it is only 100 feet (30 meters) away, the cheetah suddenly launches into the attack.

SEIZING THE PREY

After a short chase, lasting no longer than 20 seconds, the cheetah brings the zebra down. It kills it with a bite to the throat.

Moose

Gray wolves

PACK HUNTING

Gray wolves live and hunt in packs of up to 20 or more. They work together to bring down much larger prey, such as moose, often after a long chase.

EXPERT FISHERS

Grizzly or brown bears eat both plants and meat. They are very skilled at fishing. They wait in the shallows for salmon to swim upstream, leaping out of the water as they go. The bears catch their prey with either their teeth or their claws. Then they take the fish on land and carefully strip off the flesh, leaving behind the head, bones, and tail.

Grizzly bear

BEAVERS

BEAVERS are large rodents. They live in and around rivers that run through dense forests. Bark, twigs, roots, and leaves provide their food.

With their webbed hind feet and broad, flat tails, beavers are excellent swimmers. They use their huge, razor-sharp teeth to gnaw through wood. They fell trees for use in constructing their lodges.

Felling trees

Dammed lake

Dam

Underwater entrance

BUILDING A LODGE

First, beavers dam a stream to create a lake. They use trees they have felled, or build up a bank with sticks, stones, and mud. Then the beavers build their shelter, a mound of mud and branches. The chamber inside the mound is reached by several tunnels, each with an underwater entrance.

Lodge

The lake around the lodge protects the family from predators. The beavers are careful to make the lake deep enough so it does not freeze to the bottom in winter.

REINDEER

REINDEER live in the tundra, the treeless lands bordering the Arctic Ocean. In North America, they are known as caribou. They feed in herds, grazing on grasses, lichens and other small plants. The herds mostly consist of the females and their young. In winter, the reindeer scrape away the snow with their hooves to uncover the plants.

A female reindeer gives birth to one or two calves each year.

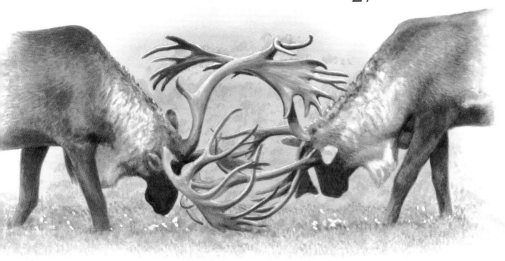

MATING

In autumn, adult males join the herds. They compete for the females by roaring or pushing each other with their antlers.

ON THE MOVE

Every winter, some reindeer herds travel south to warmer forests. This **migration** may involve a journey of more than 620 miles (1,000 kilometers). The reindeer often have to swim across wide, icy rivers to reach their destination. The hairs in their coats trap air. This helps both to keep them

warm and to float more easily in the water as they swim. Moving about in herds provides some protection from predators such as wolves.

ELEPHANTS

ELEPHANTS are the largest land animals. They are intelligent animals and may live for more than 60 years. Elephants live in family groups. After resting in the midday heat, they roam in search of leaves, shoots, and fruit to eat.

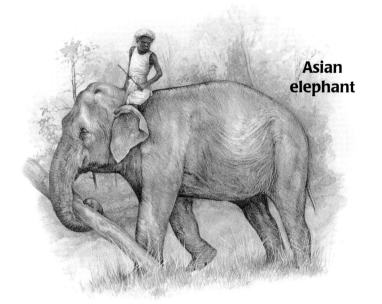

Asian elephant

AFRICAN OR ASIAN?

Elephants are found in both Africa and Asia. The African elephant has larger ears and longer tusks than its smaller relative, the Asian elephant. The Asian's back has a more humped shape than the African. Its trunk has only one lip, while the African has two. Many Asian elephants have been tamed and used for carrying heavy loads.

African elephant

Elephants drink by sucking up water in their trunk and squirting it into their mouths. They also spray water over their backs to keep cool.

An elephant can even use its trunk as a snorkel. This allows it to breathe while it swims, or when it walks on a river bottom.

USING ITS TRUNK

An elephant uses its trunk to
pick off leaves or gather up
fruits and put them in its
mouth. The trunk is powerful
enough to lift logs. But it is
also very sensitive. The finger-
like lips at the tip of the
trunk can pick up very small
objects (above). The trunk is
also used to suck up water
and as a dust sprayer. Other
uses include sniffing, greeting,
and stroking.

APES AND MONKEYS

APES and monkeys, along with humans, belong to the group of mammals known as **primates**. Most primates are good tree climbers. They have forward-facing eyes and large brains.

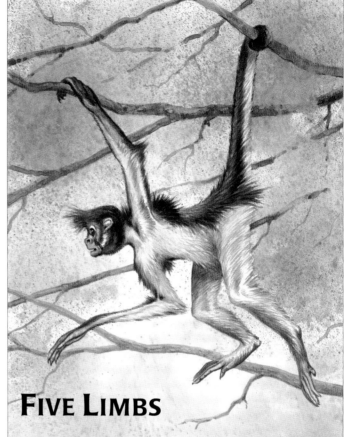

FIVE LIMBS

The spider monkey uses its long tail as a fifth limb to grasp branches as it swings through the rain-forest trees.

The orangutan's hands have long fingers and short thumbs. The ape uses them to grasp branches as it swings from tree to tree in search of its favorite fruit.

TOOLMAKERS

Chimpanzees live in African rain forests and woodlands. They spend most of their time on the ground. They eat fruits, nuts, and insects. Sometimes, they prey on other animals, such as monkeys and deer. Chimps live together in groups and defend their territory against other groups. Grooming one another helps to strengthen bonds between the chimps. Highly intelligent, they use sticks and stones as tools.

A chimpanzee pushes a twig into a termites' nest. Then it pulls it out, covered with termites to eat.

GLOSSARY

amphibians: Animals that can live both on land and in water.

estuaries: The wide part of rivers where they near the sea.

marsupial: Animals that carry their young in a pouch on their stomach, such as an opossum.

microscopic: Something so small that it can only be seen through a microscope.

migration: Movement of animals from one region to another for food or breeding.

predators: Animals that hunt and kill other animals for food.

primates: Higher mammals such as monkeys, apes, and human beings.

reptiles: Cold-blooded, usually egg-laying animals such as lizards or turtles.

INDEX